THE HARVESTING RAINWATER

DOMINATE YOUR HOME'S DROUGHT WITH THE 8 PRINCIPLES | DESIGN PASSIVE AND ACTIVE SYSTEMS AND CONTROL GRAY WATER FLOW WITH THE POWER OF NATURE ON YOUR SIDE

BRAD ALLEN

BRAD ALLEN

HARVESTING
RAINWATER

**DOMINATE YOUR HOME'S DROUGHT WITH THE 8 PRINCIPLES
DESIGN PASSIVE AND ACTIVE SYSTEMS AND CONTROL GRAY
WATER FLOW WITH THE POWER OF NATURE ON YOUR SIDE**

Contents

Introduction

Rainwater harvesting has been practiced since ancient times. Water management is an important component of sustainability and a source of ecosystem security. Rainwater harvesting is a water management tool that links human activities with natural cycles by harnessing the water that falls from the sky. Rainwater harvesting is the storage of rainwater for future use and may be used to supplement water supplies from rivers, lakes, or groundwater. If done correctly, rainwater harvesting can provide enough water for domestic and agricultural uses such as drinking water, livestock watering, irrigation, and water for house gardens.

This book shows you how to create an oasis in your yard or on your property by working in harmony with natural ecosystems rather than against them. I have included 8 simple steps to follow if you want your oasis.

The book contains a step-by-step guide to setting up a rainwater harvesting system. The book enlists the help of photographs, drawings, and line-type instructions. You'll learn how to choose the right type of system, what to include in a system and how to design it. This book is a guide to help you make sure that even the smallest quantity of rainwater is captured and the benefits become a reality. This is a way of living, implementing the simple truths of nature and nature itself.

Before you begin harvesting rainwater, it's best to check with your state departments regarding the laws governing rainwater collection in your area. Also, you will be investing in a more sustainable way of life. Some savers do not even pay for the water. Others receive a small amount of money for the rainwater for each house. Some may receive a small amount of money for their water as well. Some states even pay a small amount as a monthly allowance. There are more states than one that requires the construction of a system. This will make you think about the costs and benefits of collecting privately owned system water for public use.

1. Reduction

Reduction in water use in homes, swimming pools, and businesses.

2. Recycling

Water is recycled and reused.

3. Home Improvement

Roof decks and patios can look less cluttered with rainwater harvesting equipment.

4. Preparedness

You are ready for business interruptions or emergencies by having a sound water supply.

5. Cleaner Air

Air pollution also contributes to acid rain and the deterioration of natural resources.

6. Ecosystems

Energy is generated through the use of solar power.

7. Potable Water

Rainwater harvesting systems can be used to supply drinking water unless they are not treated by a drinking water filter system.

8. Sustainability

Rainwater harvesting systems are an important step to sustainable living. They are an integral part of living in harmony with nature.

This book is for anyone who wants to make a sustainable home, especially in relaxing areas where the dwelling is based on natural resources other than water systems. The system made in this book is an adaption of old-established strategies to help people worldwide be more sustainable. It is a way to enjoy rainwater, the gift of a plentiful water supply for life. The more you practice water management in a group effort, the more abundant life becomes. We can live with less and influence the world to do so.

1

Rainwater-Harvesting Components

Collecting rainwater means that you are taking a step closer towards self-sufficiency. An obvious incentive for you to collect rainwater is that it's free and that it reduces your water bill. Dry areas with depleted groundwater sources are perfect for rainwater harvesting; even if you live in a water-rich environment, you can conserve water through rain-harvesting. By not collecting rainwater, you are allowing this precious resource to go to waste and to flow into the ground or down the road into the storm-water drains. Not only can we use rainwater for showering, washing, cleaning and watering the garden, but it can also be used for your water heater, swimming pool or for fire fighting. Back in the day we were able to use rainwater for drinking, but these days it is advisable to test a few samples first.

Depending on where you live, the two main issues with rainwater are contaminants and water quality. The rain droplets move with the clouds through our atmosphere and cover great distances. On this journey, they come into contact with various chemicals and pollutants. Chemicals can be simple particles coming from chemical plants, paper mills, incinerators, power plants or even more seriously, radioactive particles from nuclear plants. Sadly, the day has come that acid rain, a scenario typical of science fiction movies, has become a reality of daily life. Besides our atmosphere, the rain also has to fall and collect on various surfaces. These surfaces (normally rooftops) can be coated in chemicals or might simply just be dirty from bird droppings, chimney runoff, mold, etc.

Rainwater harvesting systems come in various shapes and sizes depending on the physical area available to collect and store the rain and its intended use. While some of these systems collect rainwater directly, most collect off surfaces or rooftops.

• Collecting tanks, cisterns, and sumps: These devices can be used in various ways ranging from a simple container to gather rainwater to larger branched tanks such as the Solvay rainwater harvesting system.

• Roofs: These devices, similar to the collecting tanks, can be used in various ways, ranging from the rain-screened house that allows for tremendous drainage to a simple roof box for easier collection.

• Rain gardens and container gardens

• Infiltration Systems: To make the system more efficient and perhaps decrease its head pressure on a storage tank, some systems use filters in siphon tubes to add water to the storage tank.

• Ground storage tanks: One of the simplest and most common storage tanks is the gravity-fed system using air tanks to mix with water in underground wells.

• Irrigation systems: For those homeowners who choose to irrigate their gardens, it is extremely important to use only water sources without contaminants such as chemicals or anything that could be harmful to plants. By regulating the irrigation system, one can ensure only the amount of water needed for each plant reaches them.

In the past, it was thought that rainwater harvesting devices were mainly designed to serve as a supplementary water source, but in fact, they offer many other functions. It is also important to note that each device can be adapted to different needs of the environment, whether it be the home's temperature, humidity, and/or needs. Nowadays, with the advancement in human knowledge, harvesting rainwater efficiently with new technologies produced and created in factories is becoming possible.

The first of these technologies is the mechanical filtration system. The system comprises a filtering tank, a steady drip filter, purification coils, and a chemical additive to neutralize contamination.

Plants are living beings, and they use a lot of water in the process of their life cycle. The only source of water to meet their needs and to make them thrive comes from the water cycle. Water expands when it is absorbed by the soil. The water cools, absorbs the available nutrients, evaporates or condenses depending on the temperature, and then goes back to the cycle.

Also, the so-called "direct deposition" process is used to separate contaminants from rainwater. This is usually done by directing rainwater first through tankers before entering a sump or tank for further processing.

Another way is to simply identify the source of the contaminants (if it is water from a river, for example) and implement a system to treat the water before using it.

Rainwater can be used for both domestic and agricultural uses. Both these uses being interrelated, it is important to consider what crops we'd like to grow and how much water is needed.

Let us first perform a comparative analysis by examining the two separate paths a raindrop takes through a traditional centralized treatment facility versus an on-site rainwater harvesting system.

Traditional Centralized Treatment Facility - A Hard Path!

A centralized treatment facility consists of an extensive distribution/conveyance infrastructure of pumps, mains/laterals, treatment systems, storage reservoirs, and a comprehensive service delivery system.

Once a raindrop hits the Earth's surface, it finds its way into one of the numerous sources that municipalities pull from to meet their customers' needs. Examples of these sources are human-made and natural reservoirs, lakes, rivers, wells, and aquifers.

The raindrop is transferred into a reservoir(s) via a municipal water system's complex infrastructure and eventually to their customers from the source.

H2O is a highly effective solvent. It has ample opportunity in the complex maze of a public water system infrastructure to meet a host of foreign contaminants and absorb their harmful constituents.

Common Harmful Constituents:

- Lead
- Organic compounds (sometimes volatile organic compounds or "VOCs")
- Trash
- Animal (including human) waste (and their pathogens, e.g., e-coli bacteria)
- Industrial waste
- Pharmaceuticals
- Oils
- Chemicals

It is amazing, with all the opportunities for water to encounter so many pollutants in a public water system, that there are not more incidences of contaminated water reaching end users.

One of the main ways the public water system combats delivering contaminated water to your house is chemical treatment. There is much controversy around the chemical treatment of water, its purity, and its health impacts.

Additionally, there are justifiable concerns with the security, health, and sustainability of public water systems as water sources diminish and the population grows.

Rainwater Harvesting - The Path of Least Resistance

Regardless of the purpose, use, or types of water collected, all rainwater harvesting systems comprise four main components. These components are *catchment surface, conveyance systems (collection & distribution), storage reservoirs, and end-use delivery systems.*

Here are a couple of diagrams of potential rainwater harvesting systems. Many people start with a simple layout that includes a first flush diverter, downspout, barrel and a drip line as shown in left side of the diagram below. The other components shown here are for diagrammatic purposes to convey other possible RWH layouts that are possible.

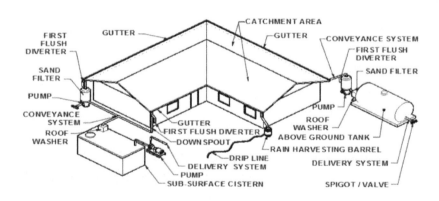

The Three Basic Components

Catchment Surface

In a rainwater harvesting system, water is collected on a catchment surface such as a rooftop or as surface water and then is transferred via the conveyance system (e.g., gutters, downspouts, etc.) to the storage reservoir(s) and thence to the delivery system for distribution.

Although rainwater harvesting systems can include human-made and natural landscaped features as catchment surfaces, I will focus on traditional residential roof catchment areas in my examples for simplicity purposes.

As far as the roofing material for water collection is concerned, there is no singular recommendation; therefore, many materials are appropriate. However, if you plan to use the water for drinking purposes (potable), the water will need to be filtered, which is covered.You must also ensure that a potable system's catchment surface is immaculate and free of contaminants.

7

Here are several catchment area materials listed in their order of preference based on "smoothness" or their ability to move water efficiently:

- Glass, i.e., Solar panels, which are gaining popularity.
- Metal: Galvalume is a commonly used material for metal roofing. It is a coated sheet metal that has a 55% to 45% aluminum-to-zinc alloy blend.
- Concrete/Clay (terracotta) Tiles: A less expensive choice, but as much as 10% loss can occur due to evaporation, porous surface texture, and inefficient flow. Also, take caution to treat the porous surface with a non-toxic sealant.
- Slate: Smooth and excellent choice for potable (drinkable) use, but it is generally more expensive.
- Composite shingles/asphalt: Can also have a loss rate as high as 10% due to rough surface texture, inefficient flow, and evaporation. The leaching of contaminates renders this choice an inappropriate choice for portable systems.
- Other: green roofs, gravel & wood shingles. Although these types are not typical, contaminates' leaching also renders these choices inappropriate for potable (drinkable) systems.

As we can see, the best surface areas are smooth ones, such as glass and metal, which contribute minimal debris to the collected water. Asphalt and composite roofs are the worst offenders. They contribute particulates and foreign objects that attach to their rough and irregular surfaces, then transferred into the storage reservoir.

Additionally, the catchment area needs to be of sufficient size to collect adequate water for your needs. In some cases, when a site does not have a big enough catchment area, a rain 'barn' or shed will be put up to increase catchment surface area.

Conveyance/Delivery System

Rainwater collected from the catchment area needs to be transferred to the storage reservoir by the conveyance system, formed from gutters that attach at the edges of a roof, leading to downspouts that terminate at your storage. The delivery system continues from the reservoir's output side and can be as simple as a dripline or a header pipe with distribution pump(s). The pumps' number and size needed to depend on the amount of water you need and the height and distance you wish to move the water.

Typically, metal or plastic (recycled?) are the preferred materials for this system. To ensure its proper functionality, the design of the guttering system is fundamental. Here are some pros and cons to considerate:

- Metal:
- Pro: lasts longer/durable/seamless/aesthetically pleasing
- Con: Costs more and can eventually rust and release contaminates into the water
- Plastic:
- Pro: Inexpensive/commonly used/easy to install/flexible
- Con: Cracks/wears earlier/non-seamless (connections points can leak) and can be less aesthetically pleasing than metal

The conveyance system, most typically gutters, should have a slope (sometimes called pitch) of ¼" for every 10' of a linear gutter to guarantee adequate water movement. If the slope is too shallow, water may stagnate in the catchment area, which will not only contaminate all the collected water, but it will also become a breeding ground for mosquitos, flies, and other unwanted pests.

If you live in an area that experiences heavy, high-intensity rainfalls, then you can expect the rainwater to shoot over the gutters. However, this may lead to a reduction in the collection, so it is best to use gutter splash/spill guards to address this issue, especially at roof valleys.

As a rule of thumb, a well-designed and efficient conveyance system should be able to transfer as much as 75%+ of the water collected from the catchment area to the reservoir(s). If you need to know precisely how much water you are harvesting, use a residential flow meter, which can come with remote monitoring options. If the catchment area is clean, completely free of debris, made from the recommended listed materials, with an adequate, clean, clog, and leak-free conveyance system, you should come as close to 75% as you physically can get.

Storage Reservoirs

The specific location that collected water is stored is known as the storage reservoir. End users access the collected water from this storage reservoir via spigots/taps or valves and, in some cases, pumps (known as the delivery system). There are numerous components in a system, but the reservoir(s) is one of the most critical and expensive parts to be carefully selected.

The storage reservoir's size, capacity, and strength depend heavily on your location, budget, and water needs. With an intelligent design, you can accurately calculate the ideal size of the reservoir you need to meet your water demands so that you do not overbuild and unnecessarily increase your budget.

For instance, if you are implementing rainwater harvesting on a small scale, you can conveniently use vessels such as barrels to store the collected water. On the other hand, if you plan to store enough water for most or all your domestic needs, then you will need to install a tank above or a tank or cistern below the ground. Note you will need to consider how you will get the water out of the underground storage which is typically done with a pump. For domestic usage, tank capacities range from 8,000 to 12,000 gallons (30 to 45+ cubic meters). However, for commercial purposes, tanks of higher capacities from 13,000 to 25,000+ gallons (50 to 100 cubic meters) are standard.

A simple way to correctly calculate the size of a reservoir is as follows:

First, calculate the annual potential cubic ft. of

Catchnment Area (Sq Ft.)		Average Rainfall (Inches)		Months		Cubic Ft. of water in a year
2500	x	55	÷	12	=	11,458

capturable rainwater.

Note: 55" is approximately the average rainfall in Florida (54.5") rounded up here for simplification.

Catchment Area: A house that measures 50'x50' has an approximate roof catchment area of 2500 Square feet.

Annual Rainfall: The annual rainfall of RWHS installation site: 55."

Gallons in a Cubic ft		Cubic Ft. of Water Annually		Gallons of Water Annually
7.5	x	11,458	=	85,938

Then, convert cubic ft. of water to gallons:

Therefore, the gross potential for collection is 85,938 gallons annually for a 50x50 sq. ft. catchment area.

Collection Efficiency

A catchment area's collection efficiency is affected by evaporation, leaks, roof type, and tree cover. However, a well-designed rainwater harvesting system can capture 75-85% of total rainfall.

Surge Point

A system surge point (the highest rainfall in a season) is the most critical element to include in your calculations when sizing a reservoir.

A simple google search of "Wettest month in *Fill in Your Area*" will help you figure out the month with the highest rainfall for your area.

Harvestable gallons projected for surge points often exceed the user demand. Instead of installing a large reservoir that is only full a few months out of the year, I recommend that you go with a smaller reservoir. Use drip systems and surface holding ponds for irrigation/landscaping or smaller, less expensive reservoirs like rain barrels.

Factors to consider

Efficiency of a system depend on many factors. Cost, climatic conditions, technical feasibility, and financial factors play a crucial role in determining if rainwater harvesting will work for you. They also help you choose the right type of rainwater harvesting system (RWHS) that will best meet your needs.

Environmental Factors

I know this may sound like a no-brainer, but the fundamental prerequisite of rainwater harvesting is rainwater! You cannot harvest rainwater usefully if it is a rarity in your region. Therefore, environmental factors are certainly the most influential considerations. To determine the ecological feasibility of your rainwater harvesting system, analyze the following elements:

- Average rainfall in the region
- Dry period duration – crucial information for designing an RWHS.

- Other water sources – such as rivers, creeks, ponds, lakes, surface holding ponds, greywater.

As a rule, rainwater harvesting is most feasible for regions that experience rainfall of 2 inches (50mm) per month for at least six months in a year.

The ideal climate for setting up a rainwater harvesting project is a tropical climate, like Hawaii, Florida, Puerto Rico, or the Virgin Islands. Tropical climates have frequent and heavy rainfall with brief dry periods. The water collected during a wet period gets used during a dry spell. The entire U.S. west coast is wetter near the coasts but have deserts in the central and east regions, which are excellent candidates for RWH.

Technical Factors

The second most important consideration for a rainwater harvesting project is technical feasibility. Consider the following before taking the plunge:

- To install the system's storage reservoir, you must have at least **20 ft²** of available catchment area on site.
- If you plan to use a roof as the catchment area, the best is metal, concrete, asphalt, or composite shingle.

There are several additional things to consider regarding the size and capacity for a rainwater-harvesting project, such as:

- The number of users (How many people will be using the water?)
- Water consumption/usage (How much per person and how will it be used.)
- Availability of alternate sources
- Availability of materials and skilled labor

Usage and consumption are the two most important technical factors. Typically, four kinds of usage patterns exist:

Occasional

Most areas may experience regular rainfalls. However, there may be intermittent dry periods that may last only a few days. An alternative source of water can be used, such as smaller reservoirs for occasional consumption.

Intermittent

In areas where dry spells separate long rain periods, the stored water is sufficient for water demands during these spells.

Partial

In this case, rainwater harvesting is used for the entire year. Therefore, harvested water is not the only source of water.

Full

Harvested rainwater is the only source of supply water. Therefore, water management and an intelligent choice of storage reservoir are necessary to ensure that you sail through the dry spell with ease.

The other two factors, which play an essential role in determining your domestic rainwater-harvesting project's feasibility are budget and availability of materials and labor. the necessary resources:

- Typical materials of catchment areas
- Typical construction of gutters and delivery systems; Metal and PVC.
- Typical construction of storage reservoirs includes brick & mortar, re-enforced concrete, fiberglass, metal, or plastic.
- Depending on reservoir and system complexity type, you will need either a tap/spigot or a pump for removing water from the storage reservoir.

Financial Factors

Budget is undoubtedly one of the dominant factors when it comes to making decisions regarding installation. The storage reservoir will be the most expensive thing in your account. A rainwater-harvesting project will require an investment in good-quality materials and workmanship to construct it.

In most cases, a roof will function adequately as an appropriate catchment surface. Furthermore, you may be able to use the existing guttering and downspouts and simply add one or more barrels to reduce cost. However, if you are looking for a more sophisticated system, it will be more expensive.

2

Conceptualize, Design, and Implement Systems

This first step is to decide the purpose of your rainwater harvesting system (RHS). This decision will allow you to set limits for the design step that's coming next. Being clear on your *goals* and understanding what you want to achieve with your RHS will ensure your design is doing what you need it to do.

Pre-planning gives you a solid point to start from, but designing an RHS can be an iterative process. You might set goals or make choices in the pre-planning stage and then realize that you need to rethink things once you reach the design or component selection step. Maybe you will notice that you are running into some soft or hard limits or cannot achieve the goals you initially wanted. Perhaps there are limitations on what parts or contractors are available to you.

This back and forth is normal, and it is better to start pre-planning with an ideal system in mind and then see if you run into any difficulties in the design stage. At this point, you can make any necessary compromises to achieve as many of your goals as possible and revise your design accordingly. Of course, none of us has unlimited money or space, so there will always be a few compromises. However, most of us can still achieve our main harvesting goals with some creativity.

Your choices will influence how you will achieve your goals. These are entirely flexible and can easily be changed at any time in the design step.

If this is your first time building an RHS, you may have to return to this step a few times as you refine your design.

Design a Rainwater Harvesting System

When designing a rainwater harvesting system (RWHS), the most crucial consideration is to ensure that you have chosen the right size and type of storage reservoir to meet your water consumption needs.

Step 1: Analyze the Requirements

Before setting up any system, you must list your requirements. As part of the conditions, you need to state the following:

Water Demand

You can decide upon the project's scale and complexity only after knowing your family's water demands. You can estimate your approximate yearly water consumption by merely multiplying your average daily water use by the number of household members and then multiplying the sum by *365* (Days in a year). According to the United States Geological Survey (United States Department of the Interior), the average daily consumption of a Westerner is between 80-100 gallons

(300-380L)/day. Since it is impossible to know precisely what the reader's daily consumption is, we will use the average of 90 gallons (340L)/day for our example:

If the average daily water consumption of one person is 90 gallons and there are *four* people in a household, then water demand is given by: *Annual Water Demand* = (90) x (4) x (365) = 131,400 gallons/annually.

Please note that you need to use the average daily consumption value. Since daily consumption values vary from person to person and season to season, absolute values cannot be used. Additionally, water consumption is not just limited to personal use. Several other factors determine usages, such as fixture type (low-flow vs. high-flow), irrigation needs, pools/spas, cleaning, laundry, dishwashing, and other household chores. Therefore, these parameters should be kept in mind when calculating the average daily per-person water consumption. If you are on municipal water, the water/sewer bills can be a good data source.

Climatic Conditions

Therefore, you must obtain the average rainfall and household consumption to determine the size of the system. It will also allow you to estimate the amount of water available for you to collect. Thus, collecting data about your local climatic condition is the first step towards designing your system. As part of your data collection, you must collect information about your area's monthly and yearly rainfall. For instance, if the average monthly rainfall is regular – between 2 and 6 inches a month; then you do not need a high-capacity storage reservoir. Specific and comprehensive analysis of climate data can add valuable detailing to your design and reduce cost.

Catchment Area

The catchment area defines the area available to harvest rainwater, such as a roof. You calculate it by multiplying the square footage of the catchment area times the average rainfall and dividing the product by 12 months.

For example, a building that is 50'-0" by 60'-0" has a roof with a catchment area of approximately 3,000 square feet. In a location with an average rainfall of 30 inches per year, the calculation would be as follows: *(3,000) x (30) ÷ (12) = 13,750 Cubic ft annually.*

Converting to gallons: (7.5) x (13,750) = 103,125 Gallons annually.

Step 2: Design Catchment Area

The most used option for a catchment area is the rooftop. However, clean the roof regularly if you wish to use it for domestic purposes.

Other important considerations for rainwater harvesting are the area available for catchment and runoff coefficient. Typically, the runoff coefficient should be at least .75, meaning your roof must allow a runoff of 75% of the water collected on it.

Material

Runoff Coefficient

Asphalt

0.90

Galvanized

0.85

Concrete

0.85

Tile (Terra Cotta)

0.70

Aluminum

0.90

Wood

0.65

Step 3: Design Conveyance/Delivery System

The most common type of conveyance system used for rainwater harvesting is a guttering system. Typically, the delivery system consists of the drip line, hose, or piping you choose to use to deliver the water to its intended use.

Conveyance and delivery components include in addition to gutters; splash guards, leaf screens, leaf guards, first flush diverters, roof washers, pump(s), and in the case of potable (drinkable) systems, final filtration.

Gutter Systems

A typical downspout arrangement connects two 45° elbows with opposite curvature, i.e., to form an "S" shape, which allows the assembly to be securely attached to the structure to give the assembly strength and rigidly. Additionally, the downspouts need to numerous enough (At least one downspout every 20'-0") and large enough to effectively transfer the water to the reservoir and prevent the gutters' overflow. Generally, PVC, aluminum, or galvanized metal pipes are used to construct this kind of assembly.

You can choose gutters made from sheet metal or prefabricated plastic. If you decide to use plastic gutters, they tend to be the most inexpensive option available. As a precaution, gutters made from PVC age and crack more readily due to direct sunlight and can droop if not properly braced; however, they are perfect for installation under roof areas. Also, gutters are available in different shapes. Typical gutter profiles are Semi or Half Circle, O-gee, "U"-Shaped, Rectangular, Trapezoidal and V-shaped.

We are concerned with a few factors when selecting shape:

P = Perimeter *[S18]*

W = Depth

A = Area*[S19]*

D^3 = Stiffness

Flow capacity, and stiffness are attainable by three dimensionless ratios:

A/P^2 = Area Ratio

W/P = Aperture Ratio

D^3/P^3 = Stiffness Ratio

Gutter Shape

Area Ratio

Aperture Ratio

Stiffness Ratio

Semi-Circle

$A/P^2 = 0.159$

$W/P = 0.64$

$D^3/P^3 = 0.032$

"U" Shape

$A/P^2 = 0.135$

$W/P = 0.39$

$D^3/P^3 = 0.059$

Rectangular

$A/P^2 = 0.125$

$W/P = 0.50$

$D^3/P^3 = 0.016$

Trapezoidal

$A/P^2 = 0.134$

$W/P = 0.80$

$D^3/P^3 = 0.013$

"V" Shaped

$A/P^2 = 0.124$

$W/P = 0.71$

$D^3/P^3 = 0.044$

Do not get too bogged down in the math. The critical takeaway here is the higher the values, the better.

Commonly used options include:

- A suitable alternative is aluminum, which is a solid and anti-corrosive material.
- If you are looking for cheaper options, you can also consider structures made from bamboo and wood planks. However, these gutters have bacteria in their organic materials, and the quality of water collected may be unfit for potable use.
- Another cost-effective alternative for guttering is half pipes. The constructs are highly efficient because of their semi-circular shapes. They are also relatively easy to create for the do-it-yourself types.

Once you have decided on the type of gutter you wish to install, you need to ensure that the gutters' construction follows the required rainwater harvesting standards. A few things that you must consider:

- You must confirm that the structure that the gutters are attached to can support the system's weight when it is full of rainwater.
- Gutters should slope towards the downspouts ate a slope of ¼" for every 10' of a linear gutter to

guarantee good water movement. If the gutter is not draining correctly, you do not have enough slope.

- The downspouts used at the end of the gutters must be capable of carrying a minimum of 90% of the collected water.

In areas that experience heavy rainfall, one should install splash guards to maximize efficient water collection into their conveyance systems. A *splash guard* is an approximately 12 - 16 inch (30 cm) long strip of sheet metal installed in the gutter at the bottom of a roof valley and acts as a barrier to keep the rainwater from shooting over the gutter.

Splash Guard

It is essential to recognize that a well-designed gutter system can play a crucial role in increasing your home's life. It allows your walls to remain dry and prevents erosion of your home's foundation.

Leaf Screens & Leaf Guards

A leaf screen or leaf guard is simply a screen, mesh, or perforated metal insert covering the gutter's top surface to filter out debris washed down from the catchment area. Screens filter out larger debris and a high concentration of smaller debris; both are excellent first-stage filtration components for a rainwater harvesting system.

Leaf Screen & Guard

Strainer Baskets

Strainer baskets are remarkably simple, tapered cylindrical (conical) baskets with sides made of mesh or screen that fit down into the downspout through the gutter's drop outlet. Strainers act as a first- or second-stage filtration for a rainwater harvesting system. Still, clean them out regularly to prevent gutter overflow due to debris plugging downspout.

Strainer Baskets

First-Flush Diverter

First-flush devices and techniques divert the rainwater collected from first rainfall from entering the storage reservoir. Although this device is not a mandatory component of the system, its use improves water quality.

These methods are:

The first of these methods is the fixed volume method. A fixed volume container collects the first flush, and when the container becomes full, it is drained.

The second method is the manual method. The downpipe (downspout) is removed from the reservoir's inlet, and the water is automatically diverted away.

The third method is a roof washer with valve and diverter tee. A simple rule of thumb to follow for how much water is sufficient for a first flush is 10 gallons for every 1,000 square feet of the catchment area. However, you will need to use your judgment whether additional gallons are required to wash your catchment area properly. This is especially true if you live in an area subject to higher-than-normal amounts of debris accumulated on your catchment area from trees and/other sources.

First Flush Diverter

Roof Washers

A roof washer is for potable systems and ideal for systems that use distribution drip lines and smaller diameter distribution/conveyance piping that can clog easily. Roof washers remove smaller debris that gets past leaf guards/screens, basket strainers, and first-flush diverters.

Pumps and Final Filtration

Step 4: Designing a Storage Reservoir

Now that you have the catchment area and delivery system ready, the rainwater harvesting system's final component is the storage reservoir. Storage reservoirs are available in several different sizes and shapes.

This method uses the number of occupants and the required water volume; it is the best method to follow if you live in a dry area. When you use this method, the storage capacity required is calculated:

Required Storage Capacity = (Demand * Number of dry months) / 12

Where demand = Per-day Per-person Water Consumption * Number of household members * 365

The demand side approach is a simple method that gives you a rough estimate of the required storage capacity. However, it does not consider certain factors like drought years and assumes sufficient catchment areas and annual rainfall.

The second approach is called the *supply-side* approach and is the average rainfall in your area. This method aims to maximize supply and is known to give a fair assessment of your storage needs. The steps that you need to follow for this method are as follows:

Supply = R * A * Cr[S22] * 0.6

Where:

R = Mean Annual Rainfall (Averaged total amount of rainfall in a year) in inches, where 1 inch of reported/recorded rain defined as the amount of water that would accumulate on a 1 square foot area to a depth of 1 inch or 144 cubic inches

Once you are clear about the required storage capacity, you will need to choose the optimal design for your storage reservoir.

Active RainWater Harvesting System

Active RainWater Harvesting System

Many people are familiar with the idea of rainwater harvesting, but fewer are familiar with the concept of active rainwater harvesting. Active rainwater harvesting is a way of storing rainwater for use at a later date, and it has a number of benefits. Above ground and underground cisterns, ponds, and other storage receptacles can all be used to store rainwater. Once it has been stored, rainwater can be used for irrigation, fountains, ponds, animals, and a variety of domestic uses. Active rainwater harvesting has a number of advantages over traditional methods of water storage. First, it allows water to be collected from a larger surface area. Second, it is easier to control the quality of the stored water. Finally, it is easier to use the stored water for a variety of purposes. As the world becomes increasingly urbanized and water becomes more scarce, active rainwater harvesting will become an increasingly important tool for sustainable living.

An active rainwater harvesting system is a type of system that actively collects and stores rainwater for future use. The most common type of active rainwater harvesting system is a roof-mounted system. This type of system consists of a catchment area, usually a metal or plastic container, that is mounted on the roof of a building. The catchment area collects rainwater from the roof and channels it into the storage container. The water can then be used for various purposes, such as irrigation, flushing toilets, or even drinking. Active rainwater harvesting systems are becoming increasingly popular as they provide an efficient way to collect and store water, helping to reduce dependence on traditional sources of water.

Passive System

A passive rainwater harvesting system is a system that relies on gravity to collect and store rainwater. The most common type of passive rainwater harvesting system is a roof-mounted cistern. Cisterns can be made from a variety of materials, but they are typically made from concrete, plastic, or metal. Roof-mounted cisterns are generally the most efficient type of passive rainwater harvesting system, as they collect water before it has a chance to run off the roof and become contaminated with pollutants. Other types of passive rainwater harvesting systems include rain barrels and surface-level reservoirs. While these systems are less efficient than roof-mounted cisterns, they are still capable of collecting and storing large quantities of rainwater.

A rooftop rainwater harvesting system typically includes four main components

- catchment area
- gutters and downspouts

- storage tank
- pumping system.

The catchment area is the surface on which rainwater is collected. This can be a roof, patio, or any other type of hard surface. Gutters and downspouts are used to collect the water from the catchment area and direct it into the storage tank. The storage tank is typically located underground and is used to store the collected rainwater. Finally, a pumping system is used to pump the collected rainwater from the storage tank to the desired location, such as a garden or landscape.

GrayWater System

Graywaters are a type of wastewater that includes all household water except for toilet water and highly concentrated wastes such as kitchen grease. When properly managed, graywaters can be reused for landscape irrigation and other purposes, saving freshwater resources and reducing pollution. One way to reap these benefits is to install a graywater rainwater harvest system. These systems collect and store graywater for later use, providing an on-site source of irrigation water. In addition, graywaters often contain nutrients that can be beneficial for plants, making them an excellent choice for landscaping applications. Graywater harvesting systems can be designed to meet the specific needs of any site, making them a versatile tool for sustainable water management.

One way to collect rainwater is through a graywater rainwater harvest system. Graywaters are defined as "wastewaters that have been used for domestic activities such as laundry, dishwashing, and bathing"

- Graywaters make up to 50-80% of the wastewater generated from household activities.
- Reusing graywaters reduces the demand on fresh water resources and can save money on water bills.
- It can also help to reduce stress on sewer systems during periods of heavy rainfall.

There are a number of different graywater rainwater harvesting systems available, and the most appropriate system depends on the needs of the household. Some common features of graywater rainwater harvesting systems include storage tanks, filters, pumps, and irrigation lines. Storage tanks can be above or below ground, and they come in a variety of sizes. Filters are used to remove particles from the water, and pumps are used to move the water to the irrigation lines. Irrigation lines can be buried underground or run above ground. The type of system that is best for a particular household depends on factors such as the climate, the amount of rainfall, and the type of soil. A qualified installer can help to determine which system is best suited for a particular location.

Earthworks

There are a variety of ways to capture and store rainwater for use in gardens. One common method is to build earthworks, such as berms or swales, which can help to slow the flow of rainwater and allow it to seep into the ground. Another option is to construct a rainwater storage system, which can be as simple as a barrel placed beneath a downspout or as complex as an underground tank. A rainwater storage system can be an ideal way to water plants during dry periods, and it can also help to protect against flooding. Regardless of the approach taken, capturing and storing rainwater can provide many benefits for gardens.

When it comes to rainwater storage, there are a wide variety of options available to suit any garden size or budget. Below are some of the most popular types of earthworks rainwater storage systems

- Infiltration basins are shallow depressions that are designed to capture and slowly release runoff. They are often used in conjunction with other measures, such as vegetated swales, to control stormwater runoff.
- Detention ponds are large, deep pools that detain and release water at a controlled rate. This type of storage is typically used in larger gardens or commercial landscapes.
- Dry wells are small pits that collect and store runoff from impervious surfaces. They help to reduce flooding and erosion by recharging groundwater aquifers.
- Retention ponds are large basins that retain a portion of the incoming runoff for an extended period of time. They are often used in conjunction with other methods, such as infiltration galleries, to control stormwater runoff.
- Waste stabilization ponds are large storage areas that use bacteria and other organisms to break down pollutants in stormwater runoff. These ponds can provide significant treatment of stormwater before it is discharged into receiving waters.

Large outdoor system

Creating an extensive non-potable outdoor system can be more challenging; however, even if you have to hire professional help, you will be in a better place to direct these professionals if you understand your goals and realistic ways to achieve them. As the size of an RHS increases, the following issues come up.

Basic systems may only require a few hours of quick planning and design, while larger systems might require multiple design iterations and more serious thought. Realizing that you can't achieve your initial goals based on hard limits or design constraints and going back to the drawing board if you can't source the right parts within your budget is a normal part of the process if you are new to rainwater harvesting.

Any complex construction project will involve more time spent designing, compromising, and resetting expectations than assembling the pieces.

Large RHS with multiple Storage Tanks

For the most common issues that come up as you design a larger RHS, a few potential technologies or

solutions are recommended below:

Problem: Insufficient Collection Area to meet demand

Solution: If only a portion of the roof surface is used as a Collection Area, you can expand the Transfer System to incorporate the entire roof surface. This expansion may require additional piping and connections in your Transfer System.

Solution: A secondary structure, such as a nearby shed, garage, or new structure dedicated to the RHS, can be incorporated into the Transfer System. Multiple Collection Areas can feed into the same Storage Tank, but your Overflow will need to be able to accommodate the total potential inflow from all sources.

Problem: Lack of space for larger Storage Tank

Solution: Buried Storage Tanks are a standard solution for lack of space. It is unnecessary to bury the entire Storage Tank, and partially buried tanks can also save space. There are additional options for Storage Tank shapes, such as thin tanks (that fit up against exterior walls), roof-mounted tanks (which will need sign-off from an engineer), or tanks that can also be placed indoors in garages or basements.

Solution: Similar to the Basic Outdoor System described earlier, smaller Storage Tanks can be linked or chained together into a more convenient shape than a single large tank.

To keep this solution simple, ensure all Storage Tanks

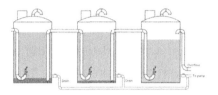

are placed at the same elevation.

Multiple Storage Tanks chained together

Problem: Fixtures are far away or require a Pump System for pressure

Solution: The most straightforward method is an all-in-one pump solution that attaches directly to or sits inside your RHS. Pump system packages designed for an RHS may be slightly more expensive but will have compatible parts, and you can install them yourself.

Solution: If getting an electrical connection is a problem, incorporating a hand pump solution is a low-cost and relatively low-maintenance option. However, you must manually pump into the pressure tank whenever you need water.

Solution: Elevating your Storage Tank can provide a small boost of additional pressure from gravity; however, every foot of elevation only gives you about 0.4 PSI. Realistically, you need about 30 feet (10 m) of height to get decent water pressure this way.

Problem: Water quality is poor, and a Treatment System is needed

Solution: A Pre-filter will increase the effectiveness of any Treatment System, and installing one should be the first step in improving water quality. The most accessible options are gutter guards and leaf screens; however, the best option is to incorporate a first-flush diversion into the Transfer System.

Solution: All-in-one Treatment System packages intended to connect directly to the RHS or set up at the fixture needing treated water are the most straightforward approach. Filtration systems can be relatively straightforward to install and maintain. Designing a chemical treatment system and determining the correct chlorine or disinfectant application rates can be difficult and time-consuming, and is not recommended for non-potable use.

Solution: Adding a slow-sand filtration system between your Transfer System and Storage Tank will ensure that water entering your tank is as high quality as possible. The downsides of this option are that these sand filtration systems may require a lot of space and have additional maintenance needs. However, the upside is that your rainwater may now be such high quality that it meets drinking water regulations and could be potable.

3
Water Harvesting Strategies

With desperate efforts to survive the increasing global rate of water scarcity coupled with water pollution, new technologies and designs are arising to solve this problem. For those seeking change, these new technologies do not guarantee a solution. They have wide open application areas with no standards to govern their design, installation, or use. There is no agreed-upon standard or regulation of water harvesting strategies.

As with any water harvesting design, the goal is a low-energy, effective system that works in moisture and climate-controlled home. The design should be simple and simple to operate. There should be no running costs, and little or no maintenance is required. It should be aesthetically pleasing and pleasing to the senses.

Passive water harvesting strategies

Passive strategies use evaporation as the primary mechanism for harvesting water. Passive strategies are the most common water harvesting strategy. Passive strategies rely largely on water evaporation due to saturation. In these strategies, the water-filled container or reservoir is kept completely dry to maximize evaporation.

Passive strategies are highly effective during dry periods. However, they are not as effective during normal weather conditions because the evaporation process slows during prolonged rain. They also require high maintenance bills because the water container must be emptied, cleaned, and dried before use.

However, the passive strategies are appropriate for medium to light rainfall or irrigation, and they are well suited for remote areas because they are not easily detectable. They are effective in both desert areas and wet areas. They can also provide domestic water supply. They can also be effective in areas that do not have access to regulated water networks.

Active water harvesting strategies

Active strategies use a flow or pressure to force water from a source to a container or reservoir. Active strategies are more complex and operate in one of three ways: at a source using a hose, over the top of the water source to a container, or in a tank.

Active strategies are best suited for areas with steady, moderate rainfall. In these areas, passive strategies are ineffective because evaporation depends on rainfall patterns and leaves the water container exposed to the sun. However, passive strategies are highly effective for short periods during years of extreme rainfall. For example, during drought years, passive strategies might be effective for a few months.

· · ·

Active strategies need to be installed and maintained, and like passive strategies, they require removing and cleaning the water container. They are also an ideal system for remote, rural areas.

Active water harvesting strategies are best installed and maintained in a specially designed tank or over the tops of fast-flowing sources (waterfalls, streams, rivers, and creeks). Tank systems need to be installed above the water source with a low-angle valve at the bottom to allow the water to feed from the source to the reservoir. Tank systems, however, are difficult to operate. They require constant maintenance and can be easily detected.

A water collector solution is a water container that drains into a non-potable water source. It can provide water for a home's needs and be used in dry areas or areas where rainwater is in short supply. The most common types are barrels, sand catchments, and leaching water pads.

Barrel systems are made up of several plastic barrels. The barrels are placed on a platform and aligned, so that the barrels fill from the ground. The filled barrels drain into an equalization tank, from which the water drains into pipes. These pipes deliver the water to the home. If a home has several water containers, they can also be linked. This makes the water delivery system more flexible so that water can be moved to different containers.

Barrel systems with multiple barrels are best located under the eves of a building. However, they can be easily seen because they are hemispherically shaped. In addition, the site should not be above a high-traffic area. These several barrels should be located on a newly constructed building so that the collection surface is level with the ground. This allows the barrels to drain quickly into an equalization tank. Placement of the barrels is also important because it must be near a water source and not on top of a roof or other area that is not level with the ground.

Sand catchments are made up of plastic bags of sand or stone. They are filled with water, then closed and left to drain. The water collects in the top part of the bag. When the plastic bag dries out, the only way to remove the water is to replace the bag. These small bags can be placed anywhere the water is needed. Sand catchments can be used for irrigation, water packaging, and storage.

A water collectors solution can also be used in home water storage. For example, a water bag could be placed in a glass container. It would be placed in direct sunlight because it would dry out quickly if it were not. In addition, the water is easily detectable.

Solar water harvesting

Solar water harvesting is a system that relies on the direct effect of sunlight on the water to produce potable water. Solar water harvesting is created by filling a reservoir with water containing a dissolved gas produced by the photovoltaic effect of sunlight. The photovoltaic effect separates the electrons from the protons causing electricity to flow. Photovoltaic water can be collected by evaporation or condensation.

A solar distillation system uses energy from the sun to make potable water. It comprises four main components: a heat source, a collector, water storage, and a processor.

The heat source can be solar radiation, electricity, gas, or a combination of two or more. In the most common form of solar distillation, the sun is used as the heat source. The device's collector absorbs light energy and is designed to capture the maximum light and solar radiation.

Based on the collector's design, harvesting water dispersed in the air is possible. The collector's design, however, determines its effectiveness in taking water from the air. A passive water collector for water harvesting absorbs the light energy that cannot be absorbed by the fluid to which it is applied. A passive water collector, therefore, has a thick wall to reflect the light. In other words, the collector is designed to absorb the light energy that cannot be absorbed by the liquid or solid against which it is applied.

Water storage is the reservoir into which the solar water is collected. It is recommended that the water storage area be larger than the collector's capacity to ensure that the reservoir will contain the water even when it collects less.

The storage tank should be black to avoid light absorption and to remain at as flat a temperature as possible. A cylindrical water container will provide a larger surface area for heat insulation and allow for a flat temperature if the container is rolled.

Water storage device and piping distribution to the use point. Before the water distribution, the water is treated as applicable to ensure it is drinkable. The natural flow of the water can be used as the system. If a fire is caused, it can be restored or reduced.

Closed-loop systems are designed to re-circulate water already used and collected. The water is reused within the home or building complex where the system is located. The closed-loop system may be installed in smaller buildings. It allows for small-scale energy conservation by requesting a balance between the energy consumed for heating and cooling.

Embankments can be used as a measure to prevent floods and can be used for open-water collection. Water collected in embankments can also be used for water collection.

Collectors can be made for the passive harvesting of water. The collector can be in the form of a well, tank, hopper, rainwater collection bucket, or other containers. However, passive water harvesting can result in longer collection times (depending on the collector's initial size) and less capacity than passive collectors. Active collectors (collectors with water flow lines) can be installed in a salt marsh, an irrigated field, or an embankment.

Certain areas have flat land, which can easily be used for roofs of buildings. If a roof is located in an area with high temperatures, the heat will make it difficult to collect water. The heat can decrease the water's capacity to produce potable water.

Plants can be placed to cool the collector. Plants cool the air around them by removing the absorbed energy and releasing the energy in the water in the form of a vapor. It cools the surface of the collector, reducing the evaporation rate.

Evaporation cools the surface of the collector, reducing the rate of evaporation.

If a solar water collection system is placed in a rural area, it could cause a decrease in the water pressure amount. The solar water collection system takes water from a small channel, which causes the water pressure to decrease. However, if the solar water collection system is large enough, it would not change the pressure of the water in the distribution system.

Water feeding lines are needed for the solar water harvesting system. A return area is also needed within the system. If the system's flow increases, a higher capacity pump is needed to keep up with the system. A valve can be installed to shut off fluid if a pump cannot be used. The fluid is kept in a reservoir when the collector is not in use. The fluid flows into the collector. The fluid then goes to the storage tank. Fluid is then pumped out of the storage tank by an electric pump.

Water is available in some way, either groundwater or surface water. Groundwater is found underground in the form of aquifers. The formation of an aquifer occurs when a section of water-bearing rock is broken. An aquifer is often created by the dissolving of rock. Rock salt dissolves along with other minerals when water is present.

Maintenance

Now that your RHS is installed and operating, keeping a regular maintenance schedule will ensure it stays that way for years or decades. Use the following checklists to design a maintenance schedule for your RHS. Maintenance checklist

Since every RHS is different, these maintenance checklists should be used as a starting point for creating your own maintenance schedule. When purchasing components, record all maintenance requirements and schedules recommended by the manufacturer in one place and use that information to adjust the checklist below to fit your system, climate, and intended use of rainwater.

Spring/Annual

- Divert Transfer System away from Storage Tank before cleaning or introducing water to your RHS
- Inspect and clean roof

- Prune overhanging branches
- Clean gutters and downspouts
- Clear any Collection Area Pre-filters
- Clear any Transfer System Area Pre-filters
- Clear any Storage Tank Pre-filters
- Check screens on Make-up and Overflow systems
- Check Storage Tank for any signs of water above the high water level
- Flush Transfer System
- Inspect Pump, reconnect if shut off for winter
- Flush Connection System
- Inspect all fittings, seals, valves, and electrical connections
- Check for moist areas indicating leaks around all connections, joints, and openings
- Ensure that no float switches are tangled with other switches or wires
- Inspect sensors, controls, and any automated systems
- Remove any dirt or debris on internal Storage Tank components
- Check sediment level in Storage Tank
- Remove sediment if more than 1/2 inch (13 mm) deep
- Check the security of the Storage Tank lock
- Change bulbs in UV Treatment System
- Inspect and clean filters in Treatment System
- Check warning lights or indicators
- Test water
- Reconnect Transfer System to Storage Tank

Summer

- Divert Transfer System away from Storage Tank before cleaning or introducing water to your RHS
- Clean gutters and downspouts
- Clear any Collection Area Pre-filters
- Clear any Transfer System Area Pre-filters
- Clear any Storage Tank Pre-filters
- Check warning lights or indicators
- Test water (if used as drinking water)
- Reconnect Transfer System to Storage Tank

Fall

- Divert Transfer System away from Storage Tank before cleaning or introducing water to your RHS
- Inspect roof surface
- Clean gutters and downspouts
- Clear any Collection Area Pre-filters
- Clear any Transfer System Area Pre-filters
- Clear any Storage Tank Pre-filters
- Inspect cold weather modifications
- Check warning lights or indicators
- Test water (if used as drinking water)
- Reconnect Transfer System to Storage Tank

Winter (Operational)

- Divert Transfer System away from Storage Tank before cleaning or introducing water to your RHS
- Clean gutters and downspouts
- Clear any Collection Area Pre-filters
- Clear any Transfer System Area Pre-filters

- Clear any Storage Tank Pre-filters
- Inspect cold weather modifications and check for ice
- Check warning lights or indicators
- Test water (if used as drinking water)
- Disconnect downspouts from the Transfer System and have them discharged to a suitable location
- Disconnect any Fixtures from the RHS and connect them to an alternate source
- Ensure all openings, inlets, and outlets are sealed to prevent contamination, insects, or animals from entering the RHS

Startup

- Ensure all openings, inlets, and outlets remained sealed while decommissioned
- Clean out disconnected pipes
- Disconnect any alternate sources and reconnect Fixtures to RHS
- Reconnect downspouts to the Transfer System
- Reconnect electrical supply to pumps, controls, and sensors
- Do a test run with a small volume of water
- Reconnect the Make-up system (if present)

Every Three Years

- Completely drain the Storage Tank, and inspect the entire RHS for damage or wear
- Clean out all sludge, sediment, or debris

- throughout the RHS

Cleaning out rain gutters

Cleaning agents

Caution should be used when selecting cleaning agents. Harsh chemicals can damage your RHS or may introduce chemical contamination that must be thoroughly removed before rainwater can safely be captured.

A solution of hydrogen peroxide mixed according to the manufacturer's instructions is sufficient for heavy annual cleaning. While bleach or other chlorine-based cleaners are often recommended, these come with their own risks, and it is not necessary to expose yourself to these dangerous chemicals. Scouring agents, heavily acidic or basic chemicals, or other cleaning agents are not recommended unless specifically identified by the manufacturer of that component as being safe to use.

Important: It can be very difficult to flush out and completely remove any chemicals you introduce into your RHS, so use caution before introducing anything new.

Additional events

If your area has seasonal sources of contaminants that can end up on your roof, you may want to schedule some additional maintenance events to deal with these. Examples could be:

- Pollen season in spring
- Trees shedding leaves or needles in fall
- Increased bird activity during migration or nesting periods
- Dust from wildfire smoke
- Nearby construction or roadwork
- Seasonal increases in insect or pest activity

Storage Tank Cleaning

In some cases, removing the accumulated sediment at the bottom of your Storage Tank may be necessary. Pumps capable of handling large debris and solids (for example, sump pumps, effluent pumps, or wet vacuums) can be used to pump out this sediment layer.

Common problems and troubleshooting

Bad smelling water

Smells are usually caused by algae growth, sediment buildup, or dead animals in the Storage Tank.

Cleaning out and disinfecting the Storage Tank will solve this problem. To prevent reoccurrence, ensure that your Pre-filter systems are installed and working correctly to reduce sediment buildup, and that there are no holes or points of entry for animals in your RHS.

Harvesting less water than expected

Blockages or leaks usually cause this situation. You can trace leaks by manually adding water to the Collection Area, but blockages can be more challenging to find. If you can't find any obvious culprits (leaves, stones, or sediment) in common blockage locations (transitions between RHS components, bends, or outlets), you may have to access cleanouts and manually add water at specific points to see where flow is being disrupted. If parts of your RHS are regularly blocked, you should alter the design or add Pre-filters at strategic locations.

Make sure to double-check your calculations for expected water capture. Check against actual values for recent precipitation and ensure you have factored in various inefficiencies like first-flush systems and the proportion of your roof used for the Collection Area.

Muddy or unclear water

If your harvested water looks brown or grayish, sediment or debris is getting into your water. Add more filters with finer screens if possible, and check that your filters are working and that sediment has not accumulated in your Storage Tank. However, filters can accumulate dirt and debris over time, and a dirty filter will eventually pass through low-quality water. A potential upgrade to resolve this could be the addition of a downpipe filter. This additional filter is placed within the Transfer System to remove debris before reaching the Storage Tank. Think of it like an advanced Pre-filter.

One possible solution is to pass the rainwater through a sand filter system before reaching the Storage Tank. This filter can significantly improve your water quality but can be complicated and may require a qualified professional for proper design and installation.

Another cause could be accumulated sediment or sludge at the bottom of your tank. During heavy rainfall, water coming in from the Transfer System could disturb this sediment layer and result in poor quality water. One option is to use a calmed or U-shaped pipe to end your Transfer System within the Storage Tank. The Transfer System pipe enters the Storage

Tank, extends near the bottom, and ends in an upwards-facing u-shape.

This modification allows rainwater to enter your Storage Tank at a slower speed so that it doesn't create

turbulence and stir up the sludge at the bottom.

Rainwater for the garden

4
Rainwater Harvesting System principles

Rainwater harvesting benefits are that it is easy, affordable, and environmentally friendly. Some rainwater harvesting systems are not even fully functional and efficient without the help of a qualified rainwater harvesting expert.

If you are considering a rainwater harvesting system or are simply interested in learning how to make rainwater harvesting systems (backflow devices) work, here is a step-by-step guide using the 8 principles of rainwater harvesting.

Principle 1 - Catch the water before it rains

Without a rain gauge to measure the flow of rain, I wonder how much we could be benefitted from detecting the incoming rainwater.

Principle 2 - Store the water in a rainwater tank

. . .

A rainwater storage tank can be any type - above ground, underground, or roof space. The designs of rainwater storage tanks are not the same. They are mostly influenced by the tank's terrain, purpose, and size.

Principle 3 - Protect the tank and a small portion of the rainwater drainage

Some water supply agencies require more protection on the tank to avoid water contamination. The tank is usually covered with a roofing material to prevent water from being contaminated by tiny creatures.

Principle 4 - Drain off gray water first

Clogged drains can be a nuisance in the household, and they could also cause health concerns. To prevent this, maintain a system that first drains gray water from the washing machine, dishwasher, and bathroom sinks.

Principle 5 - Catch rainwater from important areas first

Water from walkways, driveways, and patios is useful for irrigation and cleaning and can be used for flushing toilets and filling a cistern. Remember to ensure that the rainwater does not enter the cistern unless it has been filtered and treated so as not to contaminate the water.

Principle 6 - Organize the water flow to rainwater harvesting cisterns

There are a few ways to do this. And one way is to create a system that will direct the water flow to one water-harvesting cistern. To do so, use a water softener. It can allow water to flow from all the pipes within a house.

Principle 7 - Manage the rainwater harvesting program

Calculate the amount of water you are storing with a water-harvesting cistern and the water you consume. Have the information to know if the water-harvesting system is helping you.

Principle 8 - Use your rainwater-harvesting system to control gray water and impervious water

This principle makes it possible to use the water-harvesting system with water to manage gray water and impervious water, use the storage tank for non-potable water, and to buffer climate changes. You may now understand how rainwater harvesting systems work and what a rainwater harvesting system is. You can now make, design, and construct a rainwater harvesting system from the above list of knowledge. With this, you have the knowledge and ability to make rainwater harvesting systems, which are both easy and inexpensive.

Several examples of rainwater harvesting systems and monitoring tools are available for purchase. A few are installed by professionals, but you can build your rainwater harvesting system for a cheap and simple design.

Secrets of self-sufficiency rainwater harvesting systems

Based on the above principles, we considered two categories for rainwater harvesting systems: 'You-do-it' and 'Me-do-it.'

1. You can build a rainwater harvesting system with your current knowledge.

2. You can find a professional to help you create one.

The secret of a properly designed and installed rainwater harvesting system is to find the right combination of systems and controls. There are plenty of methods, principles, and plans available whether you are planning to build your rainwater harvesting system.

The next step would be to organize your rainwater harvesting system to help you monitor the water usage in your house. This is helpful to know how much rainwater you have collected.

The theory of rainwater harvesting has moved from a complex and mathematical one to something more common and less technical nowadays. There are several steps, practices, and guidelines that you can follow to make sure that your rainwater harvesting system is properly designed.

Benefits of rainwater harvesting systems

With rainwater harvesting, you can easily save huge amounts of money. Rainwater is abundant and is free to use, unlike water used by the government. With this, you will have a plentiful water supply, and you can use it during dry spells, saving money and time.

Rainwater collection is a must in any drought-stricken area. Rainwater harvesting helps people to store water above and underground, reducing the chance of water contamination. Rainwater has many uses, such as flushing toilets, washing clothes and cars, and irrigating gardens. As a result, a rainwater harvesting system can provide you with clean and reused water in tight circumstances.

Unless you have heard of rainwater harvesting, you are probably unaware you can use this natural resource. It will be surprising to learn that water, usually wasted, can be saved by rainwater recycling. Rainwater harvesting helps the government meet water needs by capturing the rain that falls from the sky.

Rainwater harvesting is beneficial, resulting in better health and protecting the land from deterioration. It will also help conserve precious water being wasted across the country, helping humankind.

By capturing rainwater and keeping it out of ponds, wells, and the ground, the chances of spreading diseases are reduced It keeps the water out of the ground and the reservoir and makes it possible to count the water consumed and reuse it.

People who use municipal water must consider the rainwater harvesting system. It can help the municipality to plan better by helping them to manage the utilities and public health during a long-term drought.

Rainwater harvesting is a simple, effective, and easy solution for clean drinking water. If you have no way to access potable water, rainwater harvesting is a solution. Rainwater can be used to clean and purify drinking water. Keep in mind to boil the water for safety. Rainwater harvesting is a simple solution to water shortages and is inexpensive in solving this water problem.

Rainwater harvesting is a useful way to get clean water and reduce waste. Rainwater from your roof runs off into your gutters and downspout into a storage container hidden out of sight. Rainwater harvesting can fill your swimming pool, wash your car, and even your vehicle.

Effective Treatment Methods For Harvested Rainwater Before Storage

As you construct your rainwater treatment and treated water storage facilities, there are couple of things you should know. They are:

1. The harvested rainwater must be treated before it gets to the areas of your home where it is consumed.

2. The rainwater treatment system must ensure you and your family are shielded against a range of chemical and microbiological contaminants.

Treatment Based On Point-Of-Entry (POE) Vs. Point-Of-Use (POU)

There are numerous subcategories that water treatment units can fall into. For instance, some water treatment methods are developed to improve water safety, while others are developed to improve its aesthetic qualities like taste and odour. The area of your home where water treatment takes place is also another way to classify a treatment unit.

Treatment For Microbiological Contaminants

Varieties of pathogenic microorganisms can be found in harvested rainwater. Rainwater harvested via the rooftop collection system, though typically contains very few of these microbial contaminants, still needs to be treated as though it were highly contaminated. This level of care guarantees that your family is always safe from waterborne diseases.

To nullify the health-hazard they pose, microbial contaminants can be physically removed using a filter, or rendered inactive with a disinfectant.

Irrespective of the technology you choose or the combination of technologies you utilize, you should only use certified equipment that complies with all ANSI/NSF regulations. This will guarantee that your treatment system will provide sufficient defense against pathogens.

Purifying the rain

WATER PURIFICATION PROCESS

WATER   **WATER FILTER**   CLEAN WATER

The art of purifying rainwater

R

ainwater has been called a natural resource, but rainwater is not pure water. It can contain any number of contaminants, including bacteria, lead, nitrates, pesticides, and other chemicals. In fact, rainwater can contain many different microbial species that are typically found in soil and on the ground.

Rainwater can also carry debris from the air that falls into it, and airborne particles can settle into the water. Rainwater runoff is also often contaminated by animal feces and other pollutants. Before rainwater is considered potable, it must be purified to ensure that there are no contaminants in it.

On the other hand, rainwater can be collected from roofs with gutters or from catchment areas on the ground. It is important to collect only rainwater and not runoff that has been exposed to contaminants on the ground or in a stream or river. Rainwater should be collected from one catchment area only, and it should never be combined with water from a different catchment area.

When collecting from the ground, make sure that runoff is not contaminated by runoff from roads or animal wastes. If the water source is located near a road or in an urban area, use a mesh filter to remove pollutants such as dust and debris.

For this reason, rainwater that will be stored for human consumption must undergo two stages of purification. These steps are pre-storage purification and post-storage disinfection. Pre-storage purification is used to treat and sanitize the water before it is stored for longer periods. Post-storage disinfection is used to treat the water after it has been stored in a tank or other storage vessel.

Pre-storage purification

Water that will be stored for more than a few hours should be purified to make it safe for human consumption. The process of pre-storage purification must remove all microorganisms, including bacteria, protozoa, and viruses. When water is pre-treated, it can be stored for several days until needed. Pre-storage purification should not use boiling as a method of purification. Boiling will not kill all the microorganisms in the water and can also cause some minerals, such as calcium and magnesium, to be dissolved out of the solution.

There are many other ways to purify rainwater before storage. Some methods include distillation, deionization, and reverse osmosis purification, but these processes do not remove heavy metals or chemicals. Reverse osmosis and distillation can remove some chemicals, but they also tend to remove many minerals in the water. Deionization can remove minerals as well as many chemicals and heavy metals.

Distillation works by boiling water and then condensing it to create distilled water. Distilled water is pure water that contains very few minerals and no contaminants. Distillation is considered the best method of pre-storage purification, but it does require a heat source and a container to collect the distilled water.

Pre-storage purification can also be done using chlorine to treat the water. Chlorinated water is relatively safe, and it may be used as a source of drinking water. However, chlorine cannot be added to the drinking water supply. Small amounts of chlorine are generally harmless to humans and can help clean fish and other seafood before they enter a restaurant or kitchen. However, only EPA-registered chlorine can be used for water that will be used for human consumption.

For example, Yosemite National Park has a water filtration system that provides potable water from rainwater in the park. The system was installed to provide safe drinking water for visitors and staff. The process of purification is similar to the process of purification that is done when sterilizing medical equipment or preparing food, except that it is usually performed by reverse osmosis technology.

Guidelines for selecting post-storage disinfection methods and equipment

The following are some considerations for selecting a method of post-storage disinfection:

- Does the method accomplish the goal of disinfecting the water?
- Can the method be performed with equipment that is easily transported and stored?
- Can the water be stored for an extended period while being treated?
- Will water need to be moved through pipes or a hose to ensure that it remains potable?
- Is there a way to disinfect the water without using a pump or other high-power, electric equipment?

Post-storage disinfection

Post-storage decontamination is the process of removing any pathogens and contaminants from after-stored rainwater. This is usually done by simply boiling it since boiling does not damage any microorganisms. However, boiling can destroy some minerals in the water, so it is also important to filter the water through a carbon filter before boiling.

After rainwater has been stored, it must be disinfected before it is used for drinking or other human consumption. Water that will be consumed should be treated with chlorine and then stored for at least six hours before it is used. Chlorination can also be done by adding sodium hypochlorite to the water. Sources of chlorine can include household bleach, which contains between 5% and 6% sodium hypochlorite.

Once rainwater has been treated with chlorine, it can be used for non-potable purposes such as washing dishes. Water that is consumed should be boiled or cooked before being added to other foods.

Methods of post-storage disinfection vary, and they depend on the type of water intake system that will be used. The disinfection process should include flushing the system that the water will be stored in to remove any pathogens or other contaminants that may have been added to the medium.

For instance:

- Only large particles can be eliminated by the gutter screens and prefilters. They have zero impact on pathogenic microorganisms (parasites, bacteria, viruses).
- You will have to rely solely on a disinfectant to destroy or render these pathogens inactivate, since not all bags and cartridge filters can remove them.

Compared to microfiltration and ultrafiltration membranes, reverse osmosis and nanofiltration membranes often require a greater working pressure. Larger-sized particles have the potential to clog membrane filters.

To get rid of the particles that have been stuck on the membrane surface, you must frequently flush and backwash membrane filters. Consequently, not all of the treated water will be good for consumption. Given the design, the membrane system could only be able to supply 85 to 95% of the water you harvested.

Ensure to read and abide by the manufacturer's directions. Filtration systems are generally quite durable pieces of technology. However, if you don't adhere to the manufacturer's instructions, they could easily be damaged.

Some filtration methods, particularly bag and cartridge filters, fall short of the high-performance requirements set by ANSI/NSF Standard 53. Even though some of these units have the ability to dramatically lower Cryptosporidium levels, you should always make sure that an uncertified unit has been put to the test in compliance with the specifications of the US Environmental Protection Agency, EPA, (that's if you live in the United States).

Using a filter that has been certified to match ANSI/NSF Standard 53 standards is not very important if you plan to disinfect with a high level of UV light. But the UV light cannot disinfect pathogenic microbes that are covered by other bigger particles; therefore, you must use a filter that eliminates the majority of particulate matter. It is recommended you install one of the membrane filter technologies or a cartridge filter with a pore size of 3 to 5 microns in order to maximize the effectiveness of your UV disinfection procedure.

You should only employ filtration systems that have been approved to meet ANSI/NSF Standard 61 standards to ensure that the filter does not leak unwanted particles into the water.

2. Disinfection Technologies:

There are various disinfection technologies, however some of them are more suitable for usage at home than others. For the following reasons, it is recommended you use a combination of UV light and chlorine.

- UV is incredibly powerful at destroying Cryptosporidium, however some viral pathogens need high levels of it to be eliminated. Furthermore, UV systems don't retain a residual disinfectant in your plumbing system.
- Free chlorine is almost completely ineffective against Cryptosporidium but quite powerful against viruses. Additionally, any chlorine residual in your plumbing system can be easily monitored and measured.

You might want to consider applying ozone as an alternative to UV if you don't want to keep disinfectant residues in your plumbing system. Ozone, like UV, doesn't leave a lasting residue and offers little defense against bacterial regrowth in your pipes. However, it works well against both viruses and parasites. But if you choose to apply ozone as your disinfectant, ensure to use an ozone contact vessel that has been approved in compliance with the standards of ANSI/NSF Standard 61.

3. Ultraviolet Light:

Since UV lights are more effective against parasites than they are to some viruses, it serves as a great proof of this limitation.

It's imperative to remember that a pathogen can only be rendered inactive by UV if the microbe has really been exposed to the light. Therefore, the water traveling through the UV system must be clear and free of particles in order for the system to deliver effectively. Pathogens can be rendered inactive by UV light in a matter of tenths of a second, but chlorine takes many minutes to do so.

- Chlorine can be used to disinfect water that is somewhat hazy, but UV disinfection can only work on clear water.
- Unlike chlorine, which is impacted by pH, temperature, and its concentration in the water, UV disinfection is not affected by the pH or temperature of the water.

There are liquid, solid, and gaseous forms of chlorine.

Chlorine gas has been employed as a chemical warfare agent and is exceedingly dangerous. If you don't have OSHA-approved safety gear on hand and aren't very experienced with using chlorine gas, you shouldn't use it.

For the reason that calcium hypochlorite is more stable in its solid form and comes in higher concentrations than liquid bleach, it is preferred by many homes and public water systems.

You can also use liquid bleach (sodium hypochlorite) as a source of chlorine for your water treatment. It is considered as the safest chlorine compound to use, since it is less concentrated than the other materials and is less prone to generate significant amounts of chlorine gas. However, because it is far more prone to thermal decay, you should only store enough for a 30 to 60-day period.

Similar to dissolved calcium hypochlorite solutions, sodium hypochlorite solutions are delivered using a small metering pump.

Only chlorine compounds that have been approved in compliance with ANSI/NSF Standard 60 standards should be used. This standard guarantees that the chemical will not pose any chemical or biological hazard to you or your household when used in accordance with manufacturer guidelines. This regulation also applies to any other chemical used to treat potable water.

Sadly, very few, if any, of the liquid bleaches offered by supermarkets or pool-supply companies have an NSF label. Although the NSF website contains a list of certified liquid bleaches, the majority of these manufacturers do not make bleach in small containers.

If you can't find any of the NSF-approved bleach solutions, at the very least stay away from those that have UV stabilizers or fragrances: none of such bleaches have been approved by the NSF. Additionally, avoid using the ones intended for use in swimming pools. UV stabilizers based on cyanide are frequently found in them.

5

Earthworks

Earthworks are features of the earth that have been shaped in ways that enable them to move rainwater when it reaches the ground. Earthworks are essential to survival, but they are also some of the most beautiful features of the environment.

Discovering the Ideal Shape

Many shapes of earthworks have been invented for hundreds of thousands of years, but these still fail to work effectively. It seems that there are very few solutions to this engineering problem. For example, the most common shape involves using a hard curved surface to carry the water from the sky to the ground. This is an umbrella-shaped surface with the point facing downwards. It's known as a "gutter." It works like a curved funnel, with the outlet on top. This works well but can fail because of all the water splashes. After all, the curved space is too small. The water in it can splash around too much. Also, the water can flow in many directions making a splashing pattern that doesn't easily drop to the ground.

The Perfect Earthworks

The ideal shape of earthworks depends on the shape of the land, the roughness of the ground, and the slope of the ground. It combines the magic of many curves to make a surface that will drop water to the ground without splashing a lot.

All About Earthworks

The way water gets to the ground from the sky is exciting. It starts when some rain or snow touches the surface of the earth in the form of water drops. The water drops can roll because they are spheres. They roll over the hard ground for a block or two until they reach earthwork. When the water drops reach earthwork, many things can happen.

The shape of the water drops that enter the earthwork will determine the direction of the flow. For example, a water drop with a tapered bottom will go in the direction it's pointed, but a water drop with a round bottom will go in the direction of the greatest slope.

Some water drops will follow the small ridges and bumps on the ground. Others will fall into large open areas. As a drop travels, it will let go of its water through splashes or tiny streams. This is how the water moves to the earthwork.

There are many ways that water can flow in earthwork.

(1) The water flow in earthwork can be split into two or three streams.

(2) or the flow of water in earthwork. The stream moves in one direction with the smallest surface area. Each stream follows a

different path of least resistance. It's like the path that the bright moonlight would follow in the night sky. It's a series of peals of water, one after another.

(3) The water flow in earthwork can also make many twists and turns, following the lines in the shape of circles.

When water moves through earthwork, it does so in one of three ways:

- Percolation occurs when water moves straight through the soil, like a hose.
- Seepage happens when water droplets permeate the soil and then resurface on the other side.
- Capillary action is when water moves through the soil in a circular pattern or in an S-shape.

This type of water movement can create a river-like path that carries nutrients and minerals to a specific area. All three types of water movement are essential for healthy plant growth.

When water moves through earthwork, it does so in one of three ways: percolation, seepage, or capillary action. Percolation occurs when water moves straight through the soil, like a hose. Seepage happens when water droplets permeate the soil and then resurface on the other side. Capillary action is when water moves through the soil in a circular pattern or in an S-shape. This type of water movement can create a river-like path that carries nutrients and minerals to a specific area. All three types of water movement are essential for healthy plant growth.

When water moves through earthwork, it can follow a paths that forms an "S" shape. This path can then turn and move in a complete circle around the points where it crosses itself. This shape is called a "moving system." Moving systems can be used to form a variety of shapes, including a circular circuit, triangle, or star. Each type of moving system has its own unique benefits and drawbacks. For example, a circular circuit is the most efficient type of moving system, as it requires the least amount of energy to maintain. However, it is also the most vulnerable to disruption, as any blockage in the path will cause the entire system to fail. A triangle is less efficient than a circular circuit, but it is more resistant to disruption. A star shaped system is the most complex type of moving system, but it provides the greatest resistance to disruption and is therefore the most reliable. Moving systems are an important part of many engineering projects, and understanding how they work is essential for any engineer.

When water falls on the earth's surface, it can quickly become runoff, carrying away valuable topsoil and causing erosion. One way to combat this is through the use of earthworks. By constructing terraces, ditches, and other features, farmers can redirect water flow and prevent runoff. In addition, earthworks can help to capture rainfall and store it underground, making it available to plants during dry periods. As a result, earthworks play an important role in protecting the environment and ensuring the success of agricultural operations.

Afterword

The harvested rainwater can be used for various domestic purposes like washing, flushing, and gardening. The quality of harvested rainwater is better than that of ground water, making it a good option for domestic use. In terms of environmental impact, the use of harvested rainwater reduces the demand on ground water, which often leads to over-pumping and results in lowering of water table. This also helps in recharge of groundwater aquifers. Moreover, it reduces the strain on municipal water supply system during peak demand periods. The use of harvested rainwater is therefore beneficial from both social and environmental perspectives.

A rainwater harvesting system is a great way to save water and money. The basic components of a rainwater harvesting system include a catchment area, gutters, downspouts, storage tanks, and a pump. The catchment area is usually the roof of the building, and the gutters and downspouts collect the rainwater as it runs off the roof. The storage tanks can be above or below ground, and they collect the rainwater until it is needed. The pump is used to move the water from the storage tanks to where it will be used. A rainwater harvesting system can be a simple as a single storage tank or as complex as an elaborate series of pipes and pumps.

A Rainwater Harvest system can be a great way to save money on your water bill while also helping to conserve water. However, there are a few things you can do to maximize the cost savings. First, make sure that your gutters are clean and free of debris. This will allow rainwater to flow freely into yourRainwater Harvest system. Second, consider installing a rain barrel or cistern to collect the water. This will help you to save even more water, as you won't have to rely on the municipal water supply as much. Finally, make sure to properly maintain your system. This includes regularly cleaning the gutters and ensuring that the barrel or cistern is properly sealed. By following these simple tips, you can maximize the cost savings from your Rainwater Harvest system.

Rain Harvesting Glossary

Acid rain: Acid rain is a type of precipitation that is unusually acidic. It can have harmful effects on the environment, including plants, animals, and humans. The main cause of acid rain is emissions of sulfur dioxide and nitrogen oxides from power plants and automobiles. These gases react with water vapor in the atmosphere to form sulfuric acid and nitric acid. The acids then fall to the ground in precipitation. Acid rain can damage plants, reduce the populations of fish and other aquatic animals, and erode stone buildings and monuments. In humans, it can cause respiratory problems. Acid rain is a global problem, and efforts to reduce emissions of sulfur dioxide and nitrogen oxides are important for protecting the environment.

Berm: A berm is a raised barrier, usually made of earth or stone, that is used to separate or mark off areas of land. Berms can be used for a variety of purposes, including as boundaries between properties, to create windbreaks or privacy screens, or to control runoff from rainfall. They can also be designed for aesthetic purposes, such as to create beautiful landscaping features. Whatever the purpose, berms are an important part of many landscapes. When designing a berm, it is important to consider the type of soil and the climate of the area. For example, sandy soil may require a different approach than clay soil. Additionally, berms in cold climates will need to be able to withstand heavy snowfall, while those in warm climates need to be able to resist erosion from rainfall. By taking these factors into account, you can ensure that your berm will be both functional and attractive.

Catchment surface: The catchment surface is the area of land that collects precipitation and delivers it to a particular watershed or body of water. The size of the catchment surface depends on the type of landform and the climate. For example, in a mountainous region the catchment surface is much larger than in a flat area because there is more exposed surface area for precipitation to fall on. The climate also plays a role in the size of the catchment surface. In areas with high rainfall, the catchment surface is larger because there is more precipitation to collect. The catchment surface is an important factor in determining the size of a watershed. The larger the catchment surface, the larger the watershed.

Chloramine: Chloramine is a compound formed when ammonia is added to chlorine. Commonly used as a disinfectant in water treatment, chloramine has a number of advantages over chlorine. It is more stable in water, meaning that it lasts longer and is less likely to form harmful byproducts. It is also more effective at controlling bacteria and viruses. However, chloramine can be more corrosive than chlorine, and it may cause health problems if ingested in high concentrations. As a result, water utilities must closely monitor chloramine levels to ensure that drinking water remains safe.

Chlorine: Chlorine is a chemical element with the symbol Cl and atomic number 17. It is a yellow-green gas at room temperature that evaporates readily to form a similarly colored liquid, and is present in much of the environment as a solute in seawater. As the chloride ion, it is also a major component of table salt. Chlorine has widespread uses in many different industries. It is used in PVC production, as a bleaching agent for paper, and in the chlorination of drinking water. It is also used as an antiseptic and disinfectant in many household cleaning products. In addition, chlorine plays an important role in the production of certain chemicals, such as Freon. While chlorine is an essential element in many industrial processes, it can also be dangerous if inhaled in high concentrations. For this reason, it is important to take precautions when working with this chemical.

Cistern: A cistern is a water storage tank that is used to collect and store rainwater. Cisterns can be made from a variety of materials, including concrete, plastic, and metal. They are often underground to protect the water from evaporation and contamination.

Conveyance: Conveyance is the process of transporting rainwater from the catchment area to the storage tank. A variety of methods can be used for this purpose, including gutters, pipes, and channels. The size and slope of the conveyance system will vary depending on the amount of rainfall expected in the catchment area. In order to ensure that the storage tank is filled efficiently, it is important to choose a conveyance system that is sized appropriately for the catchment area.

Countertop filtration system: A countertop filtration system is a water filtration system that is installed on or near your kitchen countertop. The system is designed to remove contaminants from your drinking water, such as lead, chlorine, and sediment. Many countertop filtration systems also improve the taste and smell of your water. The systems typically consist of a filtration unit and a separate reservoir for filtered water. Some systems also have a built-in faucet.

CPVC: CPVC is a type of plastic that is commonly used in plumbing. It is made from PVC (polyvinyl chloride) resin, and it is usually white or light gray in color. CPVC is more rigid than standard PVC, and it has a higher temperature tolerance. As a result, it is often used for hot water lines.

Downspout: A downspout is a vertical pipe that is connected to the gutter on the edge of a roof. Its purpose is to direct rainwater away from the foundation of a building and into a drainage system. Downspouts are usually made of metal, plastic, or concrete.

E.coli: Short for *Escherichia coli*, is a type of bacteria that lives in the gut of animals and humans. While most strains of E. coli are harmless, some can cause serious illness, including diarrhea, pneumonia, and even death. Symptoms of E. coli infection include abdominal cramps, diarrhea, and vomiting. If you experience these symptoms, it is important to seek medical attention immediately as E. coli can lead to serious complications if left untreated.

Earthwork: Earthwork is the process of shaping the earth to direct rainwater along the ground following the lay of the land. It is usually done with heavy equipment such as excavators and bulldozers. The most common type of earthwork is grading, which involves leveling the land to create a slope that will encourage drainage.

Gutter guard: Gutter guard is a system installed on a home's gutters to keep leaves and other debris from clogging them. Gutter guards come in a variety of materials, including plastic, metal, and mesh. Some systems cover the entire gutter while others only cover the opening at the top.

Gutter screen: A gutter screen is a device that is installed over the opening of a gutter to help prevent leaves, twigs, and other debris from clogging the gutter.

Hard water: Hard water is water that contains high levels of dissolved minerals, such as calcium and magnesium. While hard water is not harmful to your health, it can cause a number of problems around the home. For instance, hard water can make it difficult to lather soap and shampoo, and it can also leave behind mineral deposits on fixtures and appliances.

IBC tote: An IBC tote is a type of reusable industrial container used for the storage and transport of liquids, powders, and other bulk materials. IBC totes are often made from durable polyethylene or polypropylene plastic and feature a sturdy steel or aluminum frame.

In parallel: When two or more storage tanks are arranged so that water passes through them at the same time or via the same connection, they are said to be in parallel. This configuration is often used in commercial and industrial settings, where a large volume of water is required for operations. In a parallel system, each tank is filled with water as needed and the level is monitored so that all tanks remain at equal levels. This ensures a consistent supply of water, even if one tank needs to be taken offline for maintenance.

Outlet: An outlet is a conduit for water to exit a storage tank. It is typically located at the base of the tank, and its size will depend on the tank's capacity and the rate at which water is being discharged. The outlet pipe must be properly sized to ensure that the flow of water is not restricted, and it must be able to handle the weight of the stored water.

Potable: Potable water is water that is safe to drink. It contains no contaminants that could potentially be harmful to human health. While all water contains some level of impurities, potable water meets the standards set by the government for drinking water quality.

PVC: PVC is a type of plastic that is used to make pipes and tubing. It is rigid, chemically resistant, and easy to install. PVC is also less expensive than other types of pipe material, making it a popular choice for home plumbing projects.

Sludge: Sludge is a type of wastewater that is produced during the treatment of sewage. It is a thick, viscous mixture that contains a variety of solid particles, including human waste, paper, and dirt. When sewage is treated using a septic tank or an aerobic system, the solid waste settles to the bottom of the tank, where it forms a layer of sludge.

Storage tank foundation: he storage tank foundation is very important for the stability and safety of the entire tank. The foundation has to be designed to support the entire weight of the tank and the contents inside, as well as any external loads such as wind or seismic forces. The foundation also needs to be able to resist any movements of the ground, such as settlements or landslides.

Swale: A swale is a shallow ditch with sloping sides. It intercepts runoff water and directs it to a different part of your property, where it can be safely absorbed into the ground. Swales can be created using a variety of materials, including stone, gravel, and even plants. They are an effective way to manage runoff water, and they can also help to prevent soil erosion.

Ultraviolet (UV): Ultraviolet radiation is a type of energy that is invisible to the human eye. However, it can be used to safely disinfect water and render bacteria harmless. When water is exposed to UV light, the radiation disrupts the DNA of bacteria, preventing them from reproducing.

Winterization: Winterization is the process of preparing a rain storage tank for the winter months. This typically involves draining the tank and removing any foreign objects that could cause damage, such as leaves or sticks. In some cases, the tank may also be cleaned out with a hose or pressure washer. Once the tank is empty, it is important to refill it with clean water to prevent freezing and expansion.

Pump Glossary

Automatic pump: An automatic or self-acting pump is a pump that uses a small float switch to operate automatically. When the water level in the sump basin rises to a certain point, the float switch activates the pump. The pump then turns on and begins pumping water out of the sump basin and into a drain.

Booster pump: A booster pump is a device that increases the pressure of a liquid or gas. Booster pumps are used in a variety of applications, including irrigation, water treatment, and automotive systems. In most cases, booster pumps are used to raise the pressure of a fluid from one level to another. For example, a water booster pump may be used to increase the pressure of water from a municipal supply to a home or business.

Centrifugal pump: A centrifugal pump is a type of pump that uses centrifugal force to move fluids by rotation. This rotating motion creates a vacuum that sucks fluid into the center of the pump. The fluid then flows through the pump and is forced out by centrifugal force.

Deep well pump: A deep well pump is a type of water pump that is used to draw water from a deep well. It consists of a windmill-like arrangement of blades that rotate when the wind blows, creating a small amount of pressure that forces water up the pipes and into the home.

Jet pump: A jet pump is a type of water pump that uses jet propulsion to move water. Jet pumps are typically used to move water from a low-lying area, such as a well, to a higher location, such as a home or business. Jet pumps work by using a high-pressure jet of water to create a vacuum. This vacuum pulls water from the low-lying area and propels it to the higher location.

Manual pump: A manual pump is a device that uses human power to generate fluid flow. While manual pumps come in a variety of designs, they all share the common goal of using human power to generate fluid flow. One of the most common uses for manual pumps is to move water from one location to another. In agricultural settings, manual pumps can be used to irrigation systems or to remove water from ponds or other bodies of water

Priming: Priming a pump is the process of pouring fluid into the pump to remove air from the suction line of the pump. This is done to ensure that the pump will operate properly and not suck in air, which can damage the pump or cause it to fail. The priming process also allows the pump to reach its full operating potential by filling the suction line with fluid.

Submersible pump: submersible pump is a device that is completely sealed and designed to be submerged in water. It has a hermetically sealed motor that is close-coupled to the body of the pump. Junior submersible pumps are also available for depths up to 90 meters.

Made in the USA
Middletown, DE
06 January 2023